Bird Watching

Journal

Christine Dunne

Christine Dunne, Publisher

Salinas, California 2020

ISBN-978-1-7350162-1-4

Printed by Lulu Press, Inc. in the United States of America.

First Printing, 2020

Christine Dunne, Publisher

P.O. Box 2002

Salinas, California 93902

www.deadland.co

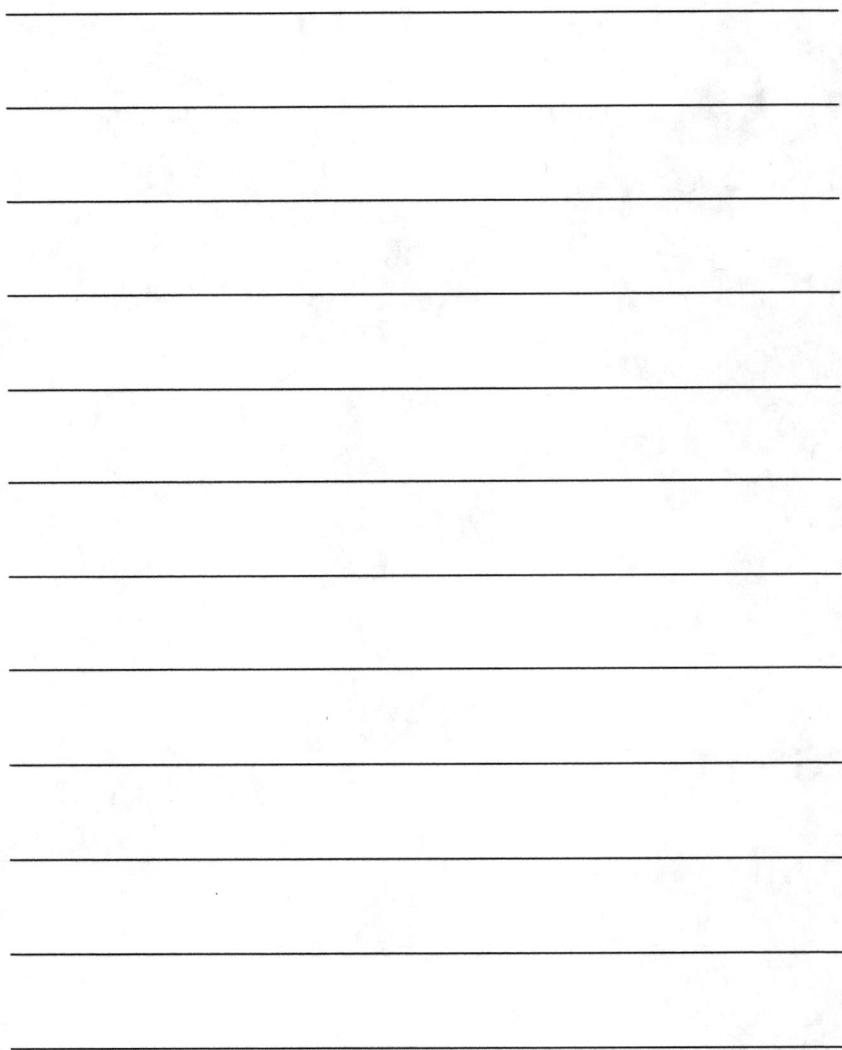

www.ingramcontent.com/pod-product-compliance
Lightning Source LLC
Chambersburg PA
CBHW070922270326
41927CB00011B/2685